The Pacific Northwest Poetry Series

Linda Bierds, Editor

THE PACIFIC NORTHWEST POETRY SERIES

This One We Call Ours

Martha Silano

LynxHousePress

SPOKANE, WA

This One We Call Ours is the twenty-fifth volume in The Pacific Northwest Poetry Series.

Cover Art: David Hytone, *Aoyf Papir (Study After the Fact),* acrylic, ink, flashe, okawara on panel.
David Hytone is represented by Greg Kucera (Seattle), Gallery 16 (San Francisco), and Studio 23 (Belgium). davidhytone.com
Author photo: Langdon Cook
Book Design: Jodi Miller-Hunter

Printed and bound in the United States of America.

Library of Congress cataloging-in-publication data may be obtained from the Library of Congress.

Lynx House Press titles are distributed by Washington State University Press, wsupress.wsu.edu

lynxhousepress.com

ISBN 978-0-89924-203-3

This book is for my parents, Marie and Alfred Silano, and for Lang, Riley, and Ruby, for without you there would be no book.

Contents

Freakishly Hot, Excessively Cold, Anticipation of Heat Dome and Wildfire Season [formerly spring]

Yearly 1,000-year Floods, 60,000 Wildfires, Fear of
Heat Dome, Bacterial Lake Closure Season
[formerly summer]

What They Said

That they'd studied four percent of the sky.
That there's this thing called the ionosphere.
That it's like looking at the sky

from the bottom of a swimming pool.
That they were counting black holes
with instruments 10x better than before,

plus algorithms, plus all these words
I didn't understand, but four percent
of the sky, but in that narrow band

to the north they found 25,000,
which means in the remaining 96%
there could be close to half a million

more. And that's what they plan to do:
count every black hole in the sky. Count
every event horizon, every instance

of swirling infinite blackness,
of the place of no return,
of the sucking power no Dyson

has ever known. The place of spaghetti-
fication. The *who knew there were so many*,
the *who knew they were visible to us*,

not just one but 25,000, each in a galaxy
like our own—substantial, giant-armed,
borne of what spilled from Hera's breast

when she realized she was nursing another
woman's child, the baby Hercules.
With its 100 thousand million stars,

with its 100 billion planets, with its one
planet with one ocean. This one we
float on, this one we call ours.

Praise beginnings; praise the end.

Joy Harjo

Carry an Inhaler, Stuck Indoors with Air Purifiers,

Air Quality Index Apps Season

[formerly autumn]

Once,

before this lake turned the color of ripened cherries, before
there was a word for *weapon* or *distance* or *phone*,
a star finished up its nucleosynthesis,

exploded its hydrogen, helium, neon and nitrogen, its sulfur
and iron, all over cosmos town. No one was around,
no one with vision or a craving for lemons.

All there was: stars and exploding stars seeding the universe
with magnesium and carbon, with graphite and diamonds.
All this, and what all else, collected into a pomegranitic

bulge that became our sun, that became the rocky planets
and the gaseous ones, that became the generous
light through pines, us and our armpit glands,

us and our *Mother, may I? No, you may not.* This was how it began,
before it cooled enough for worms and flukes, way cooler
than that instant when everything that would ever be

became, though it would be a while before figs and plumage,
rain drops and touch. But soon we had gnawing,
and soon we had fathers. Falling water

and falling in love. Before long there was work, and there was wine.
Observances like the Feast of Assumption. Soon after
there was rot and grief. But before that: electrons

and quarks. Protons and neutrons. Somehow, we got hummingbirds
and pavement, dorsal fins and cilantro. Somehow, anger
and shame and faith. Now we are a place

for lace and egrets. Now we have mouthwash and redwoods.
It's sweet like a good pear, sour like probiotic yogurt.
It began and it seems, like a novel

by Tolstoy, like it will never end, but one day—zip-zap, zap-zip—
the sun will supernova, and we will give back
our copper and plutonium, our aluminum

and titanium. The calcium in our bones will contract into dimensionless
singularity, along with all our shiny silver fillings, our stalks
of wheat, our shocks of turquoise hair.

What is beautiful? What is sad? What is apocalyptic?

The dahlia leaning over the fence like a scraggly pink mop.

The foothills at dusk, grapevines at peak production.

Ms. Kester, 52, running down the road, her dog in her arms.
Ms. Kester, cigarette dangling from her mouth: *This is not how I want to die.*

That this is the time of year to plant fall bulbs. Amaryllis. Dutch iris.
Red Velvet and African Queen.

Rescue workers digging through foundations, searching for charred bones.

The President blaming forest management.

Fire as sunset, the kind that makes you run for your camera,
share on Instagram—shades of orange and purple. Pink bursts.

That ninety minutes after it was spotted it grew, each second, by a football field.

That it was *like nighttime in daytime. Hell on Earth.*

Black Beauties, Tiger Lilies. Dance Ballerina Dance.

It looked like the sun was coming up, but we were far past sunrise.
There was this moment of Oh, that's odd … that's not good.

That some fled with important documents and mementos. That some fled
in cars that ran out of gas. That some fled their cars, ran into flames.

Retirees in mobile homes. Deeply wooded cul de sacs. Winding roads.

That in a few hours, the fire had grown to 22,000 acres. That by early evening,
it was 55,000 acres. That in one day it had traveled 17 miles.

Many stopping to jump a battery, give a pregnant woman a lift.
No one knew what effing time it was.

A friend telling me she planted one hundred stargazers.
Another putting in daffodils.

And what we saw ... at Pleasure ... what a name right now.

White specks swirling in the wind:
I wish that I could tell you it was snow.

My Watch Face

is a scary clown, a raggedy red-haired poppy.
It's tell-all o'clock and I'm striking like a grief block,
a whine shield, tuning into a radio station that will teach me

not to scream. I'm thinking if I put on
this heron pin I'll have rise vision, a reedy vibe,
mollusk-splitting hallucinations, though not the scary kind,

more like seeing my father on Lee Avenue,
raising a Rioja to his radiance, his rational-minded
attempts. My tendency? To tend to my private Euphrates,

the eddied puddles of my double helix.
October's a yellow-bellied cuckoo, opposite
of a heat-domed July. I like it this way, the drunken

orange-browns, autumn in a high ball,
shiny crows on my neighbor's black railing.
I don't mind the loss of lemonade, so long as I'll always

have the memory of poppies along a trail
to a French chateau. But enough already with deep-frying
terns and hawks. I haven't put together the side table that arrived

in 22 pieces, but I've nailed down tears,
nailed down a reason to keep asking, how could we
ever know what the first trillionth of a trillionth of a second was like?

After Apple Picking, Late Anthropocene

My puke-green stool is standing in the rain
beside a tree about to be slashed by the jaws
of a bright orange monster. There's a crate
I've filled and another crate my neighbor Taine
has filled, and we're not leaving *maybe two*

or three, we're not leaving one, and we're not
done with apple picking. Taine's gone to find
a ladder so we can reach the Northern Spy,
the Ginger Golds. Returning, she locates
the center branch – *It's grafted*, and she's right:

two varieties scarred into one—red and yellow fruits
sharing the same trunk, each with a history of its own—
one discovered in North Carolina after a hurricane
named Camille, the other hailing from the town of Peru.
I text my friend Cambria: *48 hours to pick 500 apples.*
Know a place to donate them?

Essence of progress is on the drizzly afternoon.
Intermittent downpours we don't gripe about
as we recall last summer, soot and smoke,
red sun in midday, red moon nights. N-95s.
Advisories to close our windows, stay inside.

Essence of Truffula trees and Thneeds. Of the new normal,
rising seas. I'm supposed to be grading 22 papers
but, instead, I'm coring apples for sauce, the speckled
and worm holed. Cooking them down, placing them
in sterilized jars. My daughter's in the kitchen

with her laptop, asking me to read her response
to her math teacher's lecture where he told them
the global GDP will double in the next hundred years
if we stick with gas and oil. We debate whether
she should call him a pompous bully. She is learning
Latin for *attacking the man*, she's googling synonyms

for *a-hole*, and I'm not done with apple picking—
peeling, coring, or cooking—because Juan
at City Fruit is coming by tomorrow with three crates
I'll fill for him to take to food banks, public schools.

I'm tired, wishing I could drowse, and I too can't wipe
the strangeness from my eyes when T. rex eats the roof,
metal teeth sinking into sheetrock and siding,
dragging up the viburnum, slicing the front
and back doors. I can't watch as it lifts the apple tree
by the roots, its arms like a grandma bear hugging

a grandchild. Ice? Oh yes, Mr. Frost. Not something
to study my face in, go egocentric, self-reflective,
and all about me on, but a tad more pressing.
No rest from the news of melting, no dreams
without the dream of icebergs calving. But

magnified apples *do* disappear when Juan shows up
in his pickup! No russet clarity but plenty of take
the developer's cash, leave the bank loan from the hopeful
Latinx family behind. Unlike you, Bob, I didn't want this
final harvest, would've picked 10,000 more.

After we've burned or drenched it all, after every river's
un- or over-run, after we're bulldozed under, Mother Earth
will heave a great sigh, crack open a Bud. I'm not sure
if she prefers the Tempur-flex or the Cooling Cloud
Pressure Relief foam, but it will be a long, humanless sleep.

Final Hours!

Me-n-u: we're good people, better than could
possibly extinct. Me-n-u: samey same.
A reproductive flurry ago, we'd do it
in a world all sea, all rain. Perils
were our past, but now they're increasing

like mold on a shower stall, stress and residue
induced. You know, those things like traffic,
needing to eat. Those things plastics emit.
So many by-products of our non-resourcefulness,
our live-for-today consensus, our need

to *is, is, is*. Unintended fire storms and thrice blessed
over-consumptive coming at me, at you, triggering
an earlier-than-suspected 21st Century boom, boom
pfffffzzzz. You-n-me, we know it all, are know-it-alls
because we're vertebrates, groovin' on that cortex,

at pre-fossil hegemony, tending to our evolution, unsuspecting
of our lessening apex, booted out of the Great Chain of Being
apartment complex, our Lava lights dimming. O world, late
and soon. Our habitat hatchet, our quality annihilation.
Still, something very human might save us? Undo

our finite? What we tell ourselves when the starts
of degradation germinate. When we just want
our mom, but she's sinking into the rate of our lowered
potential. Sustainable? Sequestration? Biochar? Me:
pessimist. You? The charts say mammals don't get

more than a million years, and we're at 315,000.
Okay, one-third, but that's without the hockey stick rise
on the habitat loss chart, a team without prospects.
Our human to hardly. Our seed dearth, our sperm
gone missing like the catalytic convertor

from my neighbor's Lexus. So, so long (literally), so-called
sapiens. So long, and spent well, expended massively.
It was a great genetic gyre, no? Paraphrasing our infinity,
decreasing our increase. But wasn't that cool
when our brain size quadrupled in one season? So smart,

so wedged into permanence, we thought we had longevity down,
doing that always and everywhere stomp. No one prefaced us.
We just went on panting with a broad rush, as if exeunt
were rare, as if we were on good terms with avert. Cast
our nuts in the opposite direction of imminence.
Threw caution to our planetwide decline.

Two Hundred Million Years Ago This Was All a Sea Floor,

says the vineyard owner in the Dundee Hills. He says he switched
from tequila to wine because each time his father borrowed
his Ford Bronco, he paid him back with Riesling.

The Natasha block is named for his oldest daughter. There you'll find
the Pinot reserve. Aspen rules the cuvée, Jordan the Chardonnay.
He moved up from San Bernardino for the cooler summers,

the coastal mountains, but now he says they need to irrigate, now the summers
stretch out and out in a series of 116-degree days.
We come bearing not so much a thirst

as a need for a buzz, for stories about the knock-kneed vines
planted thirty years ago, when we were knocking knees
in a booth at a pub where we first met.

The owner shuffles toward a couple sipping the Old Block Pommard,
its garnet charm, his hands trembling like his grape leaves
in the early afternoon breeze.

Last week we were in our twenties, wandering a forest, learning to distinguish
penstemon from lupine, digging out a knife from a daypack to harvest
Chicken of the Woods from the trunk of a western hemlock.

The soils of sandstone, loess, and loam make for a flinty forward,
say the tasting notes. Sandstone from the ocean, loess
from the glaciers, loam from the thirty-six times

this valley flooded when a melting glacier unblocked a massive lake.
Spans of time difficult for a human brain as we wander down
the rows of new plantings, black hoses, parched dirt.

During the Cretaceous Our Country Was Divided for Sixty Million Years

by an inland sea. The fissure was 2,000 miles long, 600 feet wide,
and 2,500 feet deep. On the left side, Laramidia.
On the right, Appalachia. Back then,

the landscape was recognizable to both sides. To both,
clams were the size of small area rugs,
turtles as big as Dodge Darts.

Both agreed *Elasmosaurus* had a streamlined body,
paddle-like limbs, a 23-foot-long neck.
If a Ginsu shark swam past,

both would laugh about how it got its name: its ability to slice
and dice. No one argued whether *Hesperornis,* a cormorant-
like bird chowed down whole by mosasaurs,

was flightless or not, whether it used its bill and teeth
to hunt down bony fish. Maybe because unity
depends on a baseline of shared reality,

all agreeing shark teeth are shark teeth, crinoid lilies
are crinoid lilies, that a fish's jaw cannot be fungible,
that you can't swap out a toothed tongue

for a smooth one. On both sides of the divide, dinosaurs roamed.
On both sides earth was earth, the landscape recognizable
to the creatures on either side.

No one on the left side and no one on the right side
rubbed their eyes and said Nope, nope,
no such thing as a plesiosaur,

no such thing as a coccolithophore. The schism was literal,
not figurative. Caused by subduction, one plate colliding
into another. At the end of the Cretaceous,

the sea dried up. Now the sea is all the middle states, those who voted red and those who voted blue. Where the sea was: the chalky remnants of a sea, of bivalves, their shiny pearls.

Today

I paddled on a lake between nature and nurture between longing for rain and a mallard taking a sort of bath a turtle basking on a dry rock that had been submerged the moon was a waning wisp the Earth frowned upon our hope our mindless emitting of CO_2 but life felt beautiful in the way a meadow or a beetle or a breeze is beautiful and I was hot and it was October and no rain and no rain and the lake not cold and I thought about tomorrow and I saw a damselfly touch down on a lily pad and a forest was burning on the west side of the Cascades what had been too wet now not too wet to burn no longer a music of sodden moss and dripping ferns but dry streambeds a deep-rooted thirst a closed road at milepost 45 a place called Wild Sky which had meant wilderness now meant unmoist tree fuel which now meant unable to see things like Mount Rainier the house up the street the bridge to the north the birds singing nothing the birds saying nothing shallow and muddy at the edge the winter migrants widgeons and mergansers not arriving the sunsets red the sunrise a red ball the big question which way is the wind blowing from the east from the north from the fires which anguish are we planting which walk to what kind of 2030 which shoulders shrugging saying it's not warming what kind of painful haze where are we and where are we going

In a Font Called Avenir Book,

I try to explain Nacho's sandpaper tongue,
how she stretches her paw along my arm.

She is gray, pale orange, warm, and already
I'm failing. Outside, the scratchy calls of jays

undulating past the fir that blocks our view
of an active volcano (it's not dormant,

its vents release steam; the inevitable landslide
that will bury Enumclaw and Sumner, its ash

once again reaching Puget Sound). The font says
shek shek shek, but it's squawkish, raspy.

Also, *keesh keesh keesh,* eliciting nostalgia
as if I'm 99, surveying my life, realizing

the best parts were quiet enough to hear the birds,
overcast afternoons when the trees stood still

and far-off voices could be heard.
Quiet enough to consider Commencement Bay,

where the Puyallup people fattened up on a glut
of salmon they could walk across, where

Simpson Tacoma Kraft erected a paper mill
requiring 40,000 dump-truck loads to cart away

its deadlier-than-Lake-Erie-when-it-caught-on-fire
sludge. To consider the Duwamish River.

My husband floats it during the odd-year humpy run;
we eat those fish because, he tells me, they never

touch bottom. PCBs, arsenic, dioxins, sewage,
and petroleum under a rayless orange sun,

a single common tansy, humble bloom that doesn't
blind, shrouded in smoke. The haze we know

like our own bodies, its vents, its faults, its uncontainable
fires quadrupling in a matter of hours, an entire region

given a red flag warning. That kind of dry fuel,
that kind of potential for being licked.

No Rain

It's October and it hasn't rained.
Why are we surprised?
Why did we think

it couldn't happen here, our home,
the Pacific Northwest? Moss,
mushrooms, mist.

The lake so low, puzzled geese.
Who wants to think about it?
It's October

and it hasn't rained. I check my phone:
Seattle 71 / Smoke. I meet a friend
who asks me to write a secret

on a scrap of paper. He does the same,
then takes out his lighter, burns
what we wrote, embers

floating above the lake, adding our smoke
to an AQI in the red, a 13,000-acre fire
seventy miles away.

The seven-day forecast: *sun, sun, part sun, part sun,*
sun, sun, sun. I don't want to ask
but I know.

The geese honk, fly south, but it's not fall
without the rain. The lake so low,
exposing a sunken forest.

It's October, time to plant spinach and kale.
We fill the watering can, dream
of picking chanterelles.

At the Wild Horses Monument in Vantage, Washington,

I pulled off the road to get closer. I thought I could drive to them,
that the exit would lead me to the top of the ridge,

but it took me to a dead end, a loop, back to the beginning.
I cut the engine, squinted up at their frozen rearing,

their stop-time galloping. Cast in (was it?)
bronze, while between me and my car a flimsy primrose,

same family as fireweed, but with tentative white petals. Also,
one I used to know, like a sickle or a question mark:

borage? But I couldn't get any closer to their spirited gaits,
to those ever-in-midair stallions, though looking back

from where I'd been there was definitely a chasm,
and looking forward? Sagebrush, a meadowlark's song,

a stiff breeze shaking the squirrel tail. To have come this close
to the memory of passing them and not stopping,

kids asleep in the back, in a hurry to see the burrowing owls
before they disappeared from their billboard perches.

Damn, is all I'm saying, because between me and them
there's a barbed wire fence, a band on the radio

called Trust, my mother who just texted me *I am a mess!!!*
Landlocked gulls and a lark sparrow, the river's

haunting columns of basalt. An empty nest a female
osprey's been arranging, big enough for a person to lie down.

When We Say It's the Little Things

we really do mean it, a friend announcing, at the end
of a long phone call, in which she shares her love
of *The Red Shoes* and *Moulin Rouge,*
 "I love you,"

how good that feels when you wake in the middle of the night,
the worth of that unexpected gesture like a lover
who cleans out the dishwasher filter without
being asked. It's the taste of that first sip

of coffee, rich and strong, the *Mr. Coffee* cup warmer
on your desk. It's having the right pen; it's a full
water bottle, a piece of orange-cranberry bread,
warm and moist, crunchy with walnuts.

We also love the big things, like Fatu and Najin, the last
two white rhinos, a mother and her daughter.
You can watch them on *YouTube*,
snorfling in the Kenyan dust,

the calming hum of insects all around them, which is more
of the little things, because mostly it's those, the insects
outweighing our human bulk seventeen times,
bush crickets and pill bugs, bot flies

and army ants that, when the time comes,
will make quick work of our little
and big bodies, our veined
and ventricled hearts.

Oh, Autolysis

All it means is self-digestion, our body's microbes doing the work
of undoing. All it means is the miracle of what had sustained us
morphing into what's needed to reduce us to nitrogen

and phosphorous, to return us to the Earth where we'll make a rich soil
for basil and thyme. The book says *leak* as cells break down,
says *rigor mortis*, says in each gut dwells

hundreds or thousands of species, which after death become
a thanatomicrobiome, from *Thanatos,* god of death,
brother of Hypnos, kin to Oizys, god of suffering,

and Moros, god of doom, the holy book of woe, the opposite
of a heart-shaped box of chocolates, the microbes,
like a rowdy mob, making their way

to the liver, the heart and brain, then everything in between.
The most hallowed parts of the body now a credenza
on which to put their feet up,

rifle through the desks of the spleen, the half-finished crossword puzzles
of the lymph nodes. I wondered how it happens, the why
of exploding abdomens, and here it is:

our bodies pre-equipped with the critters who break us down
from long-legged Vegas dancers to fodder for a cluster
of mushrooms. *Molecular death,*

the book called it. Full-on slippage into the soil.
And what do we carry? What have we carried?
Eyes that had noted the mouthlike petals

of snapdragons, the sticky white pods of milkweed,
hands that had gathered them into bouquets.
Hands that had held.

Letter to a Dead Mother

Thinking of you as I pick up flecks of oats from the kitchen floor,
put them back in the container. You know, the five-second rule.
The floor shakes as the washing machine clicks to *Spin*,

and I consider how many loads of laundry, how many times
you got down on *your* knees to mop a floor. When you left,
you took nothing. We were your *couldn't* children,

and you seemed fine with that, especially toward the end,
swallowing your *come see me's* like brambles, like bees.
Kept your dying angels, your breaking wave headlines

to yourself. If you could've had a last meal, if you'd made it past
the hallway to the kitchen that morning you took your last bath,
I bet you would've chosen leftovers: a soggy cup

of salad greens, the barley soup you'd combined with the lentils
from a few days before. Wherever you are, I hope there's plenty
of legal pads, though maybe you're writing your 15-page letters

with clouds, writing them onto the sky. And I hope they have books,
long ones by Dostoevsky, a good saga, six generations,
that it's never quiet where you are,

that there's always someone to share that joke about having 105 offspring
and not one of them comes to visit! When a comet shows up
without warning, I think of texting you *I saw it*

above the Columbia Tower! Yelled down to the kids, but it was just me
and this chunk of debris. Tonight, in your honor, it's chicken soup
with dumplings, the way your mother taught me—

egg, flour, a little salt, just enough water so the dough is sticky, not too dry.
Plop them into boiling water until they float, until they float
like an unanchored vessel on a lake with no shore.

Grief Is a Planet

~ *Camille Rankine*

Grief is a planet like Mars. The endless search for water. Finding water,
but it's frozen, deep in the Martian ground.
Is it water? Is it slush? Too salty

to desalinate? Always looking for life,
that's a constant. The same desire
for one or both,

for my parents to show up like the frozen water
deep in the soil of Mars. Searching
and listening. Water

so clear you can see trees in it, mistake the water for the trees.
Listen, it's a water ouzel on the edge of a lake.
Little dipper, dipping. Shaking out

its wings. Under. Up. Under again. My siblings *What'sApp*ing
before I'm awake: *Have you heard anything from Mom?*
I think Dad's who he would've been

if he hadn't been nuts. I tell them I've seen our parents. Not smoke,
not a mirage or bird. They flew to me, black and flapping,
two black swallows

made of something I can't place. Dissolved
in a white and swirling mist.
Who believes

we could colonize Mars, survive on a planet where storms
kick up dust that sticks to everything,
blocks out light for months.

Who wants to live on a planet where you'd have to wear a grief suit,
where it's colder than the South Pole, where the water you melt
evaporates like ghosts.

Atmospheric River, Songbird Salmonella Die-Off,

No Snow/Record-Breaking-Snow Season

[formerly winter]

December: Summer Lake

When I arrived, I wept for all the shades of brown.
Tawny. Brass. Burnt Sienna. Walnut. Buff.

The ghostly green leaves of the willows.
The black of the pier. The gray pond

with its holes like the mouths of largemouth bass.
Cattail spikes like microphones for the scaups.

I couldn't get enough of the inch or so of ice
on every twig, every blade of golden grass,

of the snow on russet hills. The chocolate mud
I glopped through, ruining my boots. Into it I sank.

January 2020

You used to have to wait for things.
Blackberries to ripen.
Cinnamon, if it arrived at all.
Tulip bulbs from Holland.

Now it's all here—asparagus
in the January dark. Papayas
and mangoes as we push
through ice and snow.

I am not against progress,
but I do not want a new disease,
something like Ebola but twice
as worse. I do not want to live

surrounded by cement and steel,
by streetlights that dim Orion's belt,
by billboards, by on-ramps
and off-ramps. I do not want

to live without the quiet mornings
of December, the astonishment
of fog, silent frost on silent
branches, without the surprise,

when I open my door, of a flicker
bobbing its head, pecking at a patch
of dirt surrounded by miles of snow.

The Problematic

of the heavy, of the atmospheric to the mountains, to the coast.
With environmental moisture. With significant intent
expected at summits north, at valleys south.

Sorta preferable to the Dome, this pre-end-of-life-
as-we-knew-it zone, otherwise known as Sunday,
projections of flooding unto Oregon,

unto Washington, unto California. It's the Weather Service, forming pew-
like plumes once known as downpours but now as rivers above,
the area of closed, of squish, of no golfing, of trees

submerged, of disintegrating roads, a half inch an hour, a spate of tweets,
an inch of snow each snowy weekend hour. Some parts of the region
devastated; other parts, no sponge but intense. Also,

something about the tropics, a rotating total, two or three active moistures,
a stream of November, and the Cascades, and the coast.
Arrives tonight, the totals not kind but in and been,

of having been in the same river twice, of some resulting moisture Monday,
displaced while reaching a stream forced to overflow, forced to issue
a warning: Lo, lo. Pressing against our watery walls

like a Great Precipitation, like a modest rivulet come to life. A blast,
but not that kind of blast. The heaviest withers (*river weather*,
December lashed). All the meteorologists said:

disconcerting. All we could muster: some joke about an ark,
an incoming potential unleashing meaning go back to bed,
meaning a bulk of terrain and a broader swath in effect,

a pineapple expressed, a recent network with pockets bulging Hawaii,
its zipper in Haiti. Narrowing to worsen, to more extreme,
roughly overall a significant spout somewhere

off the Aleutians. A day of drench, a lowland landslide, a concentrated region
of clockwise effects, channel of why and how, of more and more,
the rainiest rains, historically severe. The scale, um,

off the charts. The *are* of area, an absence of the woods that wouldn't un-inundate,
and still more inches, double-digit, record-breaking, no end
to this infrastructural impasse. The river we are,

the atmosphere we've become. The wetness taking us up and up this holiday weekend
to a bracing impact, no time for Thursday, the heftiest vapor buffeting the coast,
pounding the extra-active vulnerable. On and on and unto death, a storm,

a flooding, a posted sunrise unsunning, spinning and dumping one meter
in the already waterlogged of our largely similar highs,
said the forecasters, said the landscape of logged,

said the discussion tabled in the lake of impassible, of the latest of advisories.
Sagging washouts. Isolating stretches of flow, a basis of much dripping,
much flooding, much more dampness to fall.

Pine Siskins

All day I've been watching them dive in and out of a makeshift birdbath,
a sculpture with a depression perfect for collecting rain.
All day in the sun, high action flitting, frenetic

and alive. I reach for my binoculars, but my bare eyes are enough.
Their streaked bodies, their song going up the scale:
Zzzzrrrrrrrrrreeeeeeet!

As the sun moves from left to right, purply-pink to gray,
the birds make the most of a clear, crisp day,
crowd the perimeter,

jumping in, jumping out, perched in a circle I wish I could join.
Hope is the thing with wing bars. Hope is the thing
with a sharply pointed bill, yellow-edged wings,

a short-notched tail, though how can I omit that their population
is down, since 1970, by 80%? That *Allaboutbirds.com*
refers to them as *a common bird in steep decline.*

Salmonella outbreaks. Prey for domestic cats, poisoning from cyanide
and DDT, from pecking at asphalt and cement, ingesting motor oil,
contaminated salt. Which is why I'm surprised

when I google their name, learn this winter the United States
is experiencing a Siskin incursion, a blizzard of siskins,
yards and decks deluged. Something to do with a lack

of food up north, a southern migration in search of seeds,
which happens at *irregular intervals,* which happens,
it seems, when we need it most.

Happy Holidays from the Everglades

I give you the cardinal air plant spiking a blaze.

I give you the cloudless sulfur, the gusts and the sprawl.

I give you leaf litter and the purple iris, the mourning dove.

Morning's moisture, the dissipating fog.

I give you wind and debris, what Irma did to the Bobcat Boardwalk.

I give you the solitude of the slough, the lament of the leathery frond.

I give you the trickery of orchids. Water hemlock's poison.

I give you the phosphorous-laden swamp, the side-to-side slurp
of the roseate spoonbill.

At you I'm waving my alligator flag, my slash pine arms.

For you I'm spreading the news of the four-petal paw paw,
the snakeroot, the deltoid spurge,

the news of inundation, the steady, inescapable rise.

What Can I Tell You

that you don't already know? Maybe that our Earth
has a voice you can only hear
if you put your ear close

to the bottom of the ocean. Really. Physicists call it a hum,
an exceedingly quiet symphony, a ceaseless rumble ...
like taking a piano,

slamming all the keys at the same time, though dang,
it's completely imperceptible to the human ear,
audible only to seismometers,

if you must know. So here we are, rotating, revolving,
speeding not only through the cosmos
at 1,000 miles an hour,

but running a red light at 23rd and Pacific, late to some appointment—
eye, ear, nose, throat—speeding toward a solution
for our creaky knees, sore hips,

our agony or ennui. To have our nails trimmed, painted
Electric Geometric, to meet a friend for a drink
but first a hug,

but first a lot of hellos, a lot of we should do this more oftens,
of how did you find this place, and should we have
the Old Fashioned or the Paloma Rustica,

all the while our planet singing its little seismic song,
a ditty, like my phone alarm, on endless repeat:
Can you hear the drums, Fernando?
You were humming to yourself ...

At the Burke Museum

It was cold and rainy so we went, entered the world of phytoliths,
glass-like particles found in grass, blades leaving behind
invisible stones that persist in soil for eons. Ruby finds
a book about seeds—some lay dormant thousands of years,
some drift on ocean currents until they reach small fishing towns

like Fukushima. Black palm, hamburger bean, monkey's comb,
each pod floating in its air-filled pillow across the sea,
though I don't share what's drifting our way three years after
a 9.0 earthquake—radioactive televisions, fridges, floating docks.
Instead, we study the diorama of a swamp now busy with aircraft,

where a ground sloth rooted for tubers, grasped the branches
of willows, sparred with short-faced bears, its bones resting
12,000 years until a bulldozer smashed its skull clearing land
for a runway, the glaciers having receded, having carved the channels
of Puget Sound, the Strait of Juan de Fuca, having created

the jagged ridges of the Olympic Mountains. Eighty million years
before, ammonites, common as house flies, scooted
through shallow seas. By the time we arrive at the placard
of Pangaea breaking up, smashing into volcanic islands,
I'm woozy with the ramshackle mess of miracle, catastrophe,
impermanence—how suddenly a species can be gone.

Letter to a Post-Apocalyptic Cockroach

with apologies to Matthew Olzmann

You probably think we hated frost, rime, grout, hail, icy rivers,
icy glaciers, icy shelves keeping icy glaciers
from plunging into the sea. Probably,

you think we hated gorillas, orangutans, Edith's checkerspot, the nine-spotted
lady beetle as much as we loved Toyota Tundras, Styrofoam coolers,
Starbucks lids, the fluorocarbons in our extra-hold Aqua Net.

Knowing every minute what we were doing, metric ton by metric ton,
I bet you think we were incapable of dancing wildly in the aisles
at a what's-left-of-the-Dead show, but you'd be wrong.

Back then, when we still had the Amur leopard, a whopping total of eighty-four
because poachers killed them for their bones, steeped them in rice wine,
sold it as medicine. When we still had

Western red cedars, sword fern, Oregon grape, salal, twinflower, inside-out flower,
queen's cup, red huckleberry, and the one-sided pyrola. Man, did we ever admire
the 44,000-mile migration route of the Arctic tern. Hard to believe,

but there were seasons. Skiing. Low-lying vacation homes. Cities named Manhattan,
New Orleans, Miami. Grass died in the summer, turned emerald in September.
Do I have to tell you it wasn't all panic or worry? That some, in protest,

threw soup at famous art, but pretty much the days went on as they always had
while headlines shared the Antarctic was warming five times faster
than the global average? What were we doing,

I guess you want to know. Combusting our engines. Turning up the thermostat.
Paying eight cents, at the checkout line, for a paper bag. Buying stuff,
then donating it to the Global South, or tossing it into methane-

seething landfills. The ocean warmed 1.5 degrees, which to many seemed piddly
(most didn't know water expands as it warms). Polar ice sheets thinned.
Gale-force winds caused concrete piers to pound into each other,

break apart. Rivers fell from the sky while we fought hard for our freedom to eat beef burgers whenever we liked. And then the humans were gone, and the Earth continued to spin.

Planet killer asteroid found hiding in sun's glare may one day hit Earth

Space.com

You heard it right: a planet killer. Its name: 2022 APZ,
heading toward the rock you're spinning on.
Another bit of news, a chance

to raise your chances of having to reach for a Xanax at 2 or 3 am,
preventing that thing you often succumb to
on account of dirty bombs,

the Thwaite shelf about to drop into the Antarctic Ocean,
causing the Thwaite glacier to do the same.
What my partner refers to

as *the dark night of the soul*, which I thought he stole
from F. Scott Fitzgerald's "The Crack-Up,"
but no, it's from a poem

by St. John of the Cross. In Spanish: *La Nocha obscura del alma*,
which I kinda prefer, especially *obscura*. St. John's poem
is a narration of the journey of the soul

to the mystical union with God. Dark because, well, what else could it be?
I mean, God being unknowable, the destination obscured.
The article says there are 2,200 asteroids

that have a decent chance of hitting us, creating what they call
an Asteroid Apocalypse. Are you having a spiritual crisis
yet? I just want to say one thing, *Space.com*:

I know it's your job to share news of the cosmos, but most of us
don't need to know that in 2013 a much smaller asteroid
shattered thousands of windows

when it exploded over Chelyabinsk. That the sun hides
many more than 2,000 asteroids.
Maybe that's why

we have the eyes that we do. Maybe we're not supposed
to see this stuff. Maybe in the dark
is where we belong.

Surpassing Danger

Surpassing anything has almost always
been a lie, though the catch is the almost,

keeps ice crystals from forming on a life.
When the storm drains in reverse,

seawater gushing into the streets,
when a red moon tide is not what's

causing the seepage from underground,
when five-star service, when having it all

cannot contain the flood, when water
creeps up driveways, under security gates,

floods Lexus SUVs and Mercedes,
100s of millions will know surpassing danger

is like surpassing one's impossibles,
one's moon-jelly self. Whose dead

will float by like the bodies loosened
by Hurricane Agnes in the wake

of the Susquehanna? Not my people
or their epic poems, not my thesis

on the history of the bird, not a one
of my besties, my darlings. Not my beloved.

Gray

I have seen enough gray for one day,
in the steel-gray clouds, threatening

in the east. In the wistful trill of a sparrow,
on the floating dock where a dozen

glaucous-winged gulls preen.
I have seen enough bleached logs,

enough undersides of silver fir fronds,
enough buoys and sailboats (masts down),

enough clam shells and smooth, round stones.
I have sat long enough on a plank of wood

balanced on two stumps, gazing at a tree trunk
in the momentary sun, imagining my gray-

whiskered father in a room in a bed all alone,
his brain a cement lump, his ashen body cold.

I Am Writing a Letter

to grief. I am thinking my letter will need a stamp,
the one from when they landed on the moon.
The mail carrier will arrive

in his royal blue shorts. I will hand him my letter,
and he will hand me a small bundle
of nothing I want.

The envelope will be neither heavy nor light.
When the letter arrives, it will open
like the swirling birth of a star.

Feverfew. A heart pin made of broken seashells.
A cup of Roma. A crossword puzzle clue.
I am writing a letter

to the last star because the universe will someday
collapse. It takes a star 50,000 years to reach
adulthood, but everything dies,

including stars. It is interesting to learn
what is expected of me.
My husband says

I am taking it very well. I told a colleague
I am managing. Like when I managed
an office, answered the phone

in a fake-pleasant voice. Grief is placing its lips
on my hippocampus, that lizard part
of the brain that still hasn't

caught up with her death-rattle breath.
I am writing a letter to grief.
A framed photo of her

and my dad keeps sliding off the mantle,
which of course I'm taking as a sign.
I am writing a letter asking the mice

to keep their distance. It won't be written
in a fancy font. American Typewriter,
like her gravestone, a limestone rock.

I told a colleague I'm managing.
My letter of grief will fill
the mail carrier's sack.

Poem Written One Hundred Yards from My Mother's Grave

Of course, the sky is cloudless. Of course, when the sun sets,
we can see Gemini, four planets in a row,
Taurus's bright eye.

I'm standing on the top deck of her brother's cabin,
the one forged from cedar logs, the cabin
she never got to see,

never got to wrap herself in my Aunt Judy's red, white, and blue-checked throw,
add a log to the fire, joke about the icy outhouse toilet seat,
sing along to a battery-powered radio's

"Baby, It's Cold Outside." She's down past a grove of oaks,
or her bones anyway, in the most peaceful spot
on this 160-acre homestead.

Sitting beside her marking stone, a V-shaped cluster of Ozark sandstone,
limestone, and chert, I knew she couldn't be gone, couldn't be
in some kind of pearly-gated heaven.

Of course, when a pileated woodpecker flies close, lets go
a high-pitched *e-e-e-e-e-e-e-e-e*, I know it's her
as well as I know it's not,

but that something was making sure, when the universe formed, gravity
wasn't too strong—not enough to break atoms apart, not so weak
they couldn't hold them together, expanding into now.

Freakishly Hot, Excessively Cold, Anticipation of

Heat Dome and Wildfire Season

[formerly spring]

Vast

I was reading about exoplanets, places where there might be life,
places with open seating, place settings made of iron and clay,
where no one's heart is closed because no such thing
as hearts, & not a dry eye on this wandering orb
because they see with something else,
not stalks but more like crickets:
their viewing devices residing
in their legs. Where a bride
might witness a trillion

lightning strikes before she cuts the cake. Some are ejected
from the planetary systems in which they were formed.
Some are called nomads. Some are free-floating.
Some are rogue. It's believed the Milky Way
has billions. Exoplanets, that is.
Starless ones. Sunless ones.
Ones they call orphans.
Ones that look nothing
like ours, like us.

Chaperoning My Son's Marine Biology Class Field Trip on the 49ᵗʰ Earth Day,

where students are being asked to describe the substrate:
Sand? Cobble? Gravel? What percent? Being taught to identify what's here:
isopod, anemone, rock crab, sea brush, bull kelp. *If it's dead, don't count it*, the researcher says:

Dead crabs don't count (if they're not counting the dead,
what are they counting?). When I heard the learn'd biologist,
when I was shown the laminated list of habitat types, when I heard the teacher say

scientists aren't creative, when I asked the statistician
about the health of Puget Sound: *Octopuses are secretive.*
They know where to hide. About the sea star die-off: *They're not sure*

if it was warming seas, I too began to feel sick.
Instructed to explore, I obliged. Found a pile worm,
a Gumboot chiton, pickle and bleach weed, then wandered off

to sit on a drift log, a tree from a long-gone woods.
This is not the time to trust the Yellow-Billed Hornbill's status
is of least concern, that the Southern Blight will stop creeping north,

will stop, during times of moist warmth, pushing its way
into Wisconsin, northward with its *wide host range*, infesting
apples and strawberries, tomatoes and lettuce, and that's just one critter

to navigate—there are many more: White Peach Scale, Lanternflies,
Emerald Ash Borers, and Hemlock Wooly Adelgids. Quiet with these seniors
heading off to college because it's the world I birthed him into, because the cheetah,

black-footed ferret, and Joshua trees will soon be gone,
because that straight-up spike on the temperature chart is just as much my fault
as Standard Oil's. My doom-spewing trap is shut as one kid shouts *Get over here, you guys!*

It's a nudibranch! And there it is, on a giant rock smothered in rainbow-leaf seaweed.
I forget about moose-killing ticks, doomed Puerto Rican iguacas, skin-diseased
dolphins, the diminished leks of prairie chickens, revel in their awe.

Self-Portrait as Southern Resident Orca

For everybody *I'm speechless! Damn it, I gotta go get my camera!*
For *this must be the happiest pod.*
For you can hear them saying *there she goes again. Big one! Wow!*
For you can hear them clapping, laughing.
For I swim through the research proving there is no difference in the lifespan
of being born at Sea World or in the wild.
For 700,000 years of genetic distinction, 700,000 years of a distinct dialect evolving.
For I was misnamed *whale killer* by Spanish explorers.
For I am a dolphin.
For each year I ingest the seven million quarts of motor oil washing into the Salish Sea.
For despite being banned in 1979, each day I push through 1.5 billion pounds of PCBs.
For in my fat stores I carry your coal mining, electric appliance dependence, insecticides.
For because of you I brush against carcinogenic furans.
For I am a mother carrying her dead newborn. For I have been carrying him for days.
For thanks to my contaminated milk, he is even more toxic than I.
For you might call this behavior a tour of grief, but I have been driving my baby
to the surface so he can take a breath.
For my solitude grows scarce.
For your ships interfere with my clicks, whistles, and pulses, with knowing
where the salmon are—species, speed, size.
For the sea and I are both wide.
For the water I glide through is poisoned with viscosity index improvers; for the lapping is
laced with alkaline additives and sealants; for if you read more closely, search more carefully,
you will learn PCBs were not banned but permitted in smaller concentrations.
For I can certainly experience intense emotion.
For Monsanto's CEO makes 19 million a year but the Chemical Action Plan
lacks funding; for there is no government strong enough to save me.
For behold my spy-hopping!
For who can resist my one-syllabled, Darth Vader-like exhale?
For google *biomagnification.*
For the dusty road is my demise.
For the highway's yellow line, I die.
For I'm corralled not by my mistakes but yours.
For the doors of my duration are closing.

At the Prairie Creek Campground, Redwood National Park,

on the lawn out front by the ranger station, nine Roosevelt elk
dozing and grazing, some with their cheeks to the grass,
some sniffing the air like cats on a sunny afternoon.

My daughter held her handstand for twenty seconds,
not more than ten feet from the largest bull
I read to her from a sign: *Never walk up*
to Wild Elk. But I wasn't walking, she says.

After we goodbyed all nine, we crossed the Smith River,
descended to our camp spot with steel-bolted bear storage,
a sign insisting NO CRUMBS CAMPING.
If we saw one, we shouldn't run but act large,
fight them off with punches and kicks.

California Bay Laurel, Sitka spruce, maidenhair fern
beneath the tallest living things on Earth,
their 12-inch bark, their tannins shielding them
from insects, fungus. There had been a hundred
million acres of them. Looking up, I hated being
the same species who'd chopped even one of them down.

Just before 25 fourth graders crouched beneath a table to be instructed on the imperatives of silence and calm,

I was teaching a lesson on haiku.
Sharing a photo of a frog,
an example from Basho,

telling them they no longer needed
to count on their fingers, *5/7/5*.
Beginning to turn their attention

to the world outside the classroom,
to a cherry tree in full bloom,
asking them to watch the petals

breaking free with each small gust,
to consider what the petals resembled.
Javier waved: *Snow!*

Addison wrote *a forest is scary*.
Some were whispering.
One was confused.

One asked *Do we have to?*
Then we were quiet like the petals
falling to the ground.

Driving through Detroit, Oregon on Highway 22

We often say *things could be worse. Things could be worse*
was mostly what we said that spring of 2021:
bum knees, missed ferries, pain

at the injection sight, the fever and chills
Pfizer-induced. At every minor bend
that mantra

until we got to Detroit, the burned-out firetruck
on the side of the road, wrapped
in yellow police tape,

until we saw thousands of charred logs piled up,
waiting for truck beds, a logger at the top
of an old-growth Douglas fir

about to cut it down, working to salvage what's left.
Until we saw the rubble where homes,
where a gas station, now

a single scorched pump, until twisted, rusted metal, what might
have been a car, what had to have been
a playground, a city park.

Miraculously, a house still standing—its shockingly bright green lawn,
its twirly, cheery pink plastic flowers, its infectious hope.
How lucky, how lucky, before considering the view
from their front porch.

Leading a Nature Walk on 23rd & Yesler

I'm in favor of the foxglove,
even if parched, even if beside

a row of scraggly marigolds,
a few sad tufts of pitiful corn.

Who planted this scruffy garden?
I'm in favor of the plaintive calls

of invisible hatchlings, whisper
songs drowned out by planes,

by hip-hop blaring from a windows-
down beater teeming with teens.

Of sirens, of scooters, of the barreling
#43, of this city with its some kind

of magenta tubular, with bindweed
overtaking a hedge, with squirrel tail,

with robins cheery-upping despite
disturbance, despite asphalt,

where insects take to the air,
followed by darting swallows.

Some floozy of a blue campanulate,
each stalk blossom-crammed,

each with its pentameter stamen whirl,
each with its Fibonacci pattern of leaves,

golden ratio handed down from an Indian
mathematician, 2/1, 3/2, 5/3, 8/5,

which works as well for the flowering
of an artichoke as it does to program

a computer, a fitting continuity for a place
of engine exhaust, rampant Kentucky Blue,

where lichen clings to the bracts of leaves
like cliffs, where Day-Glo fir tips swing

in the carbon monoxide breeze. Cradling
a cottonwood leaf in my hand, the smallest

of spiders crawls out and into a world
where carbon levels soar past

400 parts per million, including this stretch
of urban corridor, where maples are busy

making samaras beside Coke cans, empty
bags of Doritos, where nectar-drunken bees

bungle through clover, where, jangling his keys,
a curious neighbor waves, says hello.

To the Ten-Lined June Beetle

Feed them to your chickens or simply squish...
—The Daily Garden

Nights you're drawn to our electric lights, to the scent
of a mate. Into the ground you go to lay your eggs,
into the ground your babies, feeding

on the roots of trees—almond, apple, cherry, plum.
You do your damage quietly, until it's too late.
But aren't we both pretty? Determined?

Industrious? But aren't you saving us
from ourselves as you destroy a crop
requiring a trillion gallons of water, three gallons per nut?

Oh, little scarab, Egyptian god of the rising sun, creator and protector,
symbol of resurrection, your lines resemble the parched ground
where pumped-dry rivers flowed.

Love Song for the Anthropocene

Always we hold them close, our phones, chaperone them
to catered parties, lug along the power strip,
the Samsonite with USB.

We have a penchant for trances. EarPods are the new boom box.
We're prone to looking down while walking,
while crossing, while surfing

and drowning, while Lyft-ing. Often, we're sipping a cold brew double dark
while scrolling. Downtime is a panther crossing I-40.
Swiping annuls our leisure. We're a hectare of text,

a harpoon of angst toward the path of entrap. We pay for the vibe
with a tilt of the wrist, touch an image of Cuvée Sauvage
and the chilled glass, speechless, sweats.

The corona of our innards glows. The weather's al dente.
The nectar's heading north and where's my phone,
it's pinging. Drought is the nightmare boss

an inch from ignite. Greed is on repeat. Earth can't talk
but it speaks. Puffins wade in waves of melted ice.
If you fancy a flood, rev your 4 x 4.

If you aren't in char, the wind will shift.
Sorry, you've missed the 8:28 –
the Bunsen burner's lit.

Denial is our teacher, stands at the front of the class
with his dandruffed 'stache. We're graduates,
like him, of the school of budworm

and stinkbug. We don't need a single degree, but soon
we'll have 2, then 3. Kim Kardashian can hire a fleet
of first responders, cruise around in her private jet,

Instagramming the Hidden Hills, but for most
the robocall rasps at 3 am, mandate to flee.
This is the ending we don't want to reach,

the chapter we don't want to read, the one where flames
engulf our ranches, our Country Livin Estates.
This is the epoch of No Plan B,

of the Speckled Wood's early emergence, before the cock's foot and Yorkshire fog
have bloomed. How this poem ends is not with a miracle,
the Hula painted frog found thriving, not extinct.

Our state of wonder will sink or be singed. Our dominance dactyl
replaced by a spate of spondees:
we are so, so fucked.

When My Brother Texts *You Guys Have a Weapon?*

I run to the side of the trail, hoard with my eyes the delicate blooms
of the bittercress, with my ears the dizzying deluge
of the Pacific Wren. May I be the lucky shopper
who snags the last case of Mount Rainier,

its lenticular shroud. When he warns me there will be shortages
(*nobody's planting anything, the meat plants will close*),
I head out to my deck, nab the clicking
of crow's feet on the railing.

When my brother texts *no need to panic, just calmly stock up,*
 I think of Rimbaud's "The Triumph of Hunger": *A taste
for eating earth and stones ... stones of churches'
crumbling gates.* Would I, if given no choice,

savor *loaves left lying in the mud?* When he texts *you might be able
to pick up a can of pepper spray or mace, a deterrent*
I'm welcoming everyone, giving
an extra-wide berth

to the woman in her eighties, her brave endeavor to walk
in a wooded park. *People could start freaking out,*
so I buy up every last carton of sky, place it
on my neighbor's porch.

Nine Billion Years After the Big Bang,

the Earth began to form when gravity pulled together swirling gas and dust.
It sounds so easy, doesn't it? Gravity plus dust equals a planet
so many of us have a soft spot for.

I wish I could tell you how we got from gravity and dust
to hoodoos and the Gobi Desert, to the Rockies
and the Mississippi gopher frog.

It's like writing an email to someone you haven't seen in forty years,
trying your best to catch them up. *So yeah, it took a while,
but it finally cooled. Oceans formed,*

and from them life emerged. No matter how much you try to cram into that missive,
you'll never capture how you got from 20 to 60,
from a roach-ridden one-room studio

in Portland to what all else ensued. Things like the composition
of the atmosphere: 78% nitrogen, 21% oxygen, 1%
a bunch of stuff—argon, CO2, and neon. *Neon.*

Also, why is the inner core's temperature 9,800 degrees? Why is the mantle
1,800 miles thick, with the consistency of caramel?
Why is the crust 19 miles deep?

We live on a sphere that's always moving at a speed of 18.5 miles per second.
Come again? Living out our days kind of oblivious, the same way
most of us can't rattle off the names of the muscles in our backs,

explain what a duodenum is for. *Splenius capitis. Teres minor and major.*
Helps to break down food. On quiet afternoons, the clouds
like in that painting "The Blue Boat" by Winslow Homer.

It doesn't seem possible, does it, that the minerals that make up our bodies
are billions of years old, that it all started with a temperature
of a thousand trillion degrees that cooled to form galaxies,

planets, and stars. Lichens and fungi. This planet where the ground doesn't seem like it could ever fissure, go goopy, swallow us all.

The Race to Save the Planet's Inhabitants

Carpets of columbine gracing our shoulders,
we thrust into the wildfire plumes.
Neonics nixing our buzz,

we dim our cynical, brighten our weevil resolve,
dive into the maelstrom. From the opposite
corner, someone shouts *Heat dome!*

We go in search, tumbling into the terminus
of the dry times, those tunnels of scald.
We eat brown shoots for breakfast,

scorched mussels for lunch. It's an orca ordeal,
an orchestrated-by-oil-burning brand
of fickle weather. We're off course,

of course, boot over bottom, horizon over horsetail,
but still we go in search of a butterfly cloud—
an orange haze of tortoiseshells,

choke down the math of yet another species
in decline. Swim to the center
of a nest that once held

three billion birds, each of us implicated,
each of us drowning. Each of us
gasping, grasping an oar.

Lunching with Frank O'Hara on My Deck

The Russian sage is sprouting leaves from a brown stem
dormant since November. Weeks ago, I pulled away
the dead, unsure whether it was a perennial,

thinking *now the phone can be answered, nobody calling,*
only an echo / all can confess to be home and waiting.
The era of the landline is over, though we've found

even better ways of returning home to pain. I don't need
to invite him over, offer, instead of yogurt,
a plate of lasagna with extra cheese:

he lands on my deck railing, carried on the wings
of a rock dove, its purple iridescence
an amazement, telling me

my lasagna's the best, just the right amount of ricotta,
reminding me of the preciousness of moss, helping
me see the blooming Pampas grass

is a clowder of cats, bushy tails waving in the April breeze.
How are you feeling he asks. Like the owner of a palace,
I confess, though more than a little distressed.

Letter to My Grandchildren

with apologies to Matthew Olzmann

Yeah, well, I'm dead. Upside-down in a bucket of not so great.
Absolutely ask me what I think of the jet stream gone berserk.
For sure, blame me and my Hyundai Elantra for the rain that falls,

when it falls, like a scorpion stings. Hey, I bet you're asking,
what up with not doing diddly to stop the North Atlantic right whales
from going extinct, for letting them get tangled in fishing gear.

Yeah, you should be pissed. We all kept quiet while planes flew over
our heads, while houses were built without mandatory solar,
while this guy with orange hair told us it was all a hoax

made up by China. We sat on our duffs sipping Montepulciano
when we could've been keeping California condors,
Mariana fruit bats, and white and northern black rhinos

from dwindling to zero. I was busy planning my next vacation,
a break from the un-euphoric work week. It was pretty cool
when there were birds—I loved when a dozen crows

would divebomb an eagle over the Interstate. I guess now only humans
use hooks, spears, and bent wire to impale their lunches. So sorry.
I absolutely probably would've done something to stop the dreadful

inevitable, but I was designing crop tops for *H&M*, running to the store
for a six-pack and a steak. Now the entire Amazon is a field of soy, home
to one-hundred million head of cattle. I harpooned your joy. I'm the ancestor

to blame for all you could have had, including the long-mustached
Emperor tamarin, and the Cape Verde great skink. Damn,
if it doesn't suck. Bees were a good thing, even when they stung.

Mirabile dictu

La Traviata reminds me my mother is a corpse.
Those first few days the birds came so close,

a yellow warbler while I ate my ham on rye.
Now, a cloud is giving the moon a black eye.

Which clothes was she buried in I do not ask.
Who touched my mother last?

I'm a being chased by birdsong: sparrow, bunting, wren.
This is neither nightmare nor game.

Though they lace my kale, I do not brain the snails.
She's underground. Surrounded by a wooden fence.

The fog is giving the night a reason to feel safe.
So sorry, little flower, for not knowing your name.

That week she became the heron in the muck;
that week a chickadee landed on our truck.

She who was not my compass, not my map.

My Son Has a Girlfriend and Other Things I Can't Tell My Mother

Like that I'm making the cranberry bread I just sent her the recipe for. That I finished *Dutch House* and guess what: Maeve dies! Can't tell her we got a refund today from our car insurance company—$66.00. She would've loved that. Or that Lang asked me if I'd be willing to watch a YouTube video on quantum physics at seven in the morning, and I said sure. Something about a cat in a box. Schrodinger's? Anyway, it sits inside a cardboard box where there's cyanide (or was it arsenic?). We don't know if the kitty will live or die, but it doesn't matter because if it dies the cat that would have lived spawns a new universe where it goes about its feline business. But does the universe where the cat goes have another cat box with a different cat in it? And does someone on that universe put cyanide (or arsenic) in the box and that cat spawns another universe, etc.? Is this cat alive/dead/cyanide (or arsenic) situation infinite? Cat in a box after cat in a box? If so, how different is this from the origin story with the world held up by a turtle, of asking *but who holds up the turtle?* and being told *it's turtles all the way down?* Like that Lang accused me of hoarding rice because I bought one extra bag. *Now someone's gonna leave the grocery store bereft*, forgetting how fast we go through rice, how QFC and Whole Foods are often out. Like I feel bad now too: who wants to be rice-less? Oh, and Mom? The video mentioned entanglements. I know how much you appreciate those. Like I was wearing my flowered balaclava thingie over a surgical mask, my glasses fogging up. It just sucks, I want to tell her; can't you just please be alive?

On the Outside

I'm a tousled cosmos, a strongly-winded daisy,
a wound-down reef. On the inside
I'm just me: heart beating
like a garbage truck backing up your street,
an ordinary Wednesday, an odd or even day
in June. What I want is difficult
to pronounce, is something I need to look up,
check the spelling (does it even exist?)
Or the simplest: an apple,
a small white bowl of green grapes, a mask that fits.
Two days before she died, my mother told me:
of all my births, yours was the easiest;
of all my births, yours was the one that did not
almost kill me. Two days before she died,
how did I let her hang up the phone
without telling me everything? Each day I wake up
wanting to ask her one last question, which bird
landed in her plum tree—
catbird, thrasher? Both male and female cardinals?
House finch, mockingbird, Carolina wren.
On the outside, a tornado passes
through my hair. On the inside, I incline toward the bird
with the thinnest whistle. Daily, I reach for my phone
to ask her for that pierogies recipe. Every day the daisy
loses another petal, another bract. On death
I'm no expert. On death, I know less than the scale
that gives it a negative infinity.

I Love the World

but now I know a mother can work in her garden for ten hours,
not know it's her last day alive. Now I know
no one's there to deadhead the zinnias

and the feverfew. Even though the world is filled
with injured geese and gulls, millions of acres
of smoldering trees,

I still love cantaloupe, how it sits on the kitchen counter
waiting for my spoon to scoop its firm and juicy flesh.
Even after I saw a photo

of my mother's casket draped with Grandmother's quilt,
I still loved hearing about the field of white daisies
down the road from her grave.

The world is both the wheat plowed under to make way for strip malls
and a sunset like spilled orange juice above a gray lake.
Joy resides in the mountains

of Styrofoam and Ziplocs, while sorrow suffuses my mother's backyard,
its cardinals and finches, its hummingbird perched in a plum tree
that lost all its branches in a terrible storm.

Yearly 1,000-year Floods, 60,000 Wildfires, Fear of

Heat Dome, Bacterial Lake Closure Season

[formerly summer]

Time and Distance

Time is like a sandal in a cave: here I was.
If no heat is exchanged, time does not exist.
The little blue macaw is gone. The distance grows.

What one needs is a helping verb, a word like *does*.
What one gets instead is a cosmic abyss.
Time is like a sandal in a cave: here I was.

There are planets no telescope will disclose.
Though no one's sure, our nature is to shift.
The little blue macaw is gone. The distance grows.

There should be more silence, less applause.
We should be more like Neptune grass. It persists.
Time is like a sandal in a cave: here I was.

Horseshoe crabs preceded the dinosaurs.
Each spring they convene for a spawning blitz.
The little blue macaw is gone. The distance grows.

If you come bearing peanuts or a cage, a crow recalls.
For just one day, let me grow a carapace.
Time is like a sandal in a cave: here I was.
The little blue macaw is gone. The distance grows.

Failed Attempt at Mythmaking

I wanted to write about terns. Which terns? The Caspians nesting
on a very hot roof. What the conservation scientist said:
They confuse the dust from the cement plant

with beach. During the heatwave the flightless chicks
jumped. Jumped. Fell into gutters, got stuck.
Some made it to the ground,

where they also got stuck. I asked a friend how to write a poem
about animals in pain. *What if you wrote it*
as a short poem? Seven lines max.

I'd been reading Sappho, had brief on the brain,
but this wasn't a fragmented lyric situation.
The newspaper said *A number*

were too injured and euthanized. A number,
but no number provided. The dead birds
were juveniles. The friend said

Maybe you could write it as an Ars poetica,
or just keep writing past the end.
The ones that lived

were treated for burns on their feet. And fluid support,
which I guess means water. A number,
however, were not.

A volunteer with Audubon said he'd never seen
birds jump from their nests from such a height,
but it had never been 106 in Seattle.

I could bring in Hephaestus, god of fire,
patron saint of craftsmen.
His workshop's inside

a volcano. I could bring in Dante's *Inferno: Welcome to the city
of woe. Welcome to everlasting sadness ... to the grave cave.
You, who have no hope ...* etc.

My friend wrote: *I think it has to do with imagining things
have agency beyond our knowledge, which they do.*
They mistake our cement piles for beach,

a tar-covered roof for a prime nesting site.
I could mention the patron saint
of animals, I could write past,

so far past—when they're not,
when we're not—
but I won't.

My Daughter Is Drawn to Blue Flax

Because its tall, airy stems bend
toward the light.

Because it overtakes hillsides
scorched by fire.

Because the flowers you see in the morning
are replaced by new ones by the next.

Because if you plant it in your garden,
butterflies will come.

Because a farmer in Pasco who planted eighty acres said his field
resembles a blue lake.

Because flax is not simply a pretty flower
but can be woven into baskets.

Because the seeds can be ground into grain, pressed into oil
that soothes and calms.

Because 30,000 years ago cave dwellers dyed, spun, and braided
its fibers into snares and seines, rope and string.

Because *linen* shares a root with *lineage.*

Because it's three times stronger than cotton.

Because every fiber of her waves in the same breeze.

You're Surrounded

by green forgiveness, summer's narcotic sorbets. It's difficult
to think about disappearing at the six o'clock hour
in late August, the unlacking light, the dinner

of sour and breakdown, a roadway of strain, the lit votive
of asking: What was a rose, a bud, a thorn?
So much of the past is like the fog

you wake to, unable to see beyond the dumpster
or the tunnel slide. Waiting for the foibles
to vamoose, the fuckups to flee.

Most of the day had been a rose: Pterosaur herons, woodpecker peel,
groves of Pacific madrones. The bud a new friend who lives
in Pennsylvania, sends missives about her dream view:

dolphins and cormorants, when her office abuts
a computer lab. When your daughter asks
but what was the thorn,

should you share the list that begins with pinkeye and croup,
ends with carrion beetles and coffin flies?
One luminosity gives way to another—

the annoying spotlight piercing your bedroom window,
though also an almost-full moon,
a small chunk missing

from the upper left, two planets you can't see but imagine rising over the harbor.
And what about your day, my dear? What vexed the traffic
of neurons in your cosmos-resembling brain?

All week taking photos, trying to capture the uncapturable, a week's worth
of two hundred million firings per second, a small fraction
of the two-three billion fired in a lifetime. Annoyed

the blinds are cheap, inferior at blocking the punishing light.
This close to the sea, surrounded
by the birthplace of tears.

Why I Love to Garden in My Front Yard

with gratitude to Ross Gay

Because I was picking jalapenos,
but now I'm biting into one of Yoshi's figs.
Because she's holding a silver bowl heavy with dozens,

saying *take*. Because I'm biting into
my first fresh fig, stopping mid-bite, eyeing
the small white strands that might be writhing.

Because they are not maggots,
because I am savoring, along with sweet flesh,
the sun, air, and sky. Yoshi in her white capris, her big straw hat.

Because Eddie smothered our Kentucky Blue parking strip
with dirt, built three raised beds, so where we used to mow
we're growing pumpkins, stalks thick as Jack's, six green, fuzzy fruits

because honeybees found their way into each starry bloom.
Because my husband can't keep pace with the fecundity, put up
fifteen pints of Dilly Beans, because the tomato plants are so heavy

with fruit the cages have given way, cannot be righted;
because the snails have finally given up on making fine lace
of our kale, and though the beets did not produce, their tops added zing

to the stir-fry. Because when I posted about it on Facebook,
several friends warned us we'd be robbed. That as of this date,
not a single serrano, eggplant, or carrot has been pilfered. Because

a neighbor stopped by to chat, said squash blossoms
are a delicacy, showed me how to dip them in a mixture
of beer and flour, fry them in oil. Because we're biting into sunshine.

Poppy Love

You're an egg yolk, a paper-thin pond
at dawn, anthers waving like anemones

in a rising tide. With your pineapple stigma,
you and all your sisters glorify the path

to the sea. Floppy hat for a sanderling,
you hail from California, land of mythology

and gold. Impossibly common, ubiquitous
along back roads, you would never hurt

an approaching katydid but serve
as a parasol for a semi-palmated plover.

You're the dab of mustard
on Van Gogh's palate, tint

he transformed into sunflowers
and wheat fields. With your bronchial

leaves I remember to breathe,
with your spear-like capsules

I'm reminded people are not
always kind but that blossoms

give way to fruiting bodies,
that your round brown seeds persist.

I Have to Deepen My Know

ledge because it's shallow like a tarn
in late August, because I don't
have a grasp on the rate of melting,

on the sponge-like Greenland firn
keeping oceans from rising.
My trifling know ledge,

unexcavated, undredged. The movement
north of commas and pikas, little egrets,
the strengthening of tropical storms

with names like Matthew and Gaston.
My ledge brims with gaseous laughter,
with buoyant conclusion, peals calamity-

rich. I will find me a walk-behind trencher,
a skid-steel loader, dig this sad excuse
for a reef into a mantle. With my significant

foxhole, I will gorge and moat, trough and dig
till I've dug to the bedrock of mastery;
I will make this ledge of mine a mountain,

more sloped, more shored, more earth worked.
I will scoop and scrape till I surface the contents
of the whale that washed up on a Spanish coast,

fifty-nine plastic items in its gut—two flowerpots,
a spray canister, 37 pounds of trash bags.
With this trailing pipe, I am tracking the moth,

the mole-like Pyrenean desman elevating
eight inches an hour. With my modish know
ledge, I will no longer possum but posit

not a wall but a walrus's need for ice—
its floating pre-school, its staging ground
for lunch. On my sturdy berm they will glide

and glissade. Congregate. Give birth.

The Prize Bird of Friday Harbor

That day when everything came down
to a covey of quail. Quail-like strutting
on the hike to Jakle's Lagoon, quail-like
squeal of my daughter climbing a sap-

smeared pine. In the morning it was quail
yoga, even when we were doing side crow.
Throughout an afternoon of the melodious
waterfalling of white-crowned sparrows,

we kept hoping to hear it: the song of a bird
with a freaking pompadour. Bird that runs
better than it flies, juts its neck forward
and back like a human imitating a fowl.

I've heard rumors people eat them.
I once had a boss cautiously share
there's a *season*, along with mourning dove,
Hungarian partridge, snipe. Over the din

of purple finch and yellowthroat, it all came down
to the waiting, that unmistakable *woo-HOO-hoo!*
Because their breasts are gold-plated.
Because their offspring follow in a crooked line.

In the Late Anthropocene

Venus fades from the eastern sky, off to grab
a morning latte, drops the lid, drops out of sight

because light pollution, because the haze.
Empty pop bottles litter the yard

of an about-to-be-demolished bungalow,
its triple-grafted apple tree—Liberty,

Empire, Ginger Gold. To get here, it had to get
way out of hand, way out of foot and spine:

greed like the Snake River at flood stage,
the town of Mora filling sandbags. *Hoping*

for the best, but what is best, but what is hope
but cramming your head into a sandbag,

dragging it down to the powerful water.
A hummingbird ticks and buzzes, sizing up

the shriveled *Salvia* while neighbors amble by
with their Heelers and Corgis, talking teardowns

and buildouts, where each new house will rise.
How many stories they want to know,

how high will it go, how many stories
of the scorching, the drowned.

After Several Sharp Caws in Succession

Something about orcas not having a complex language.
Something about a scale, our words at the top.

Something if they're so smart why don't they skip
the Chinook, pig out on Coho and chum.

Something about picking blackberries without Galway Kinnell,
his *stalks very prickly*, a line I've been repeating since 1984.

Something about CO2, each year seven million more metric tons.
Something called the Alberta Tar Sands, where 50,000 acres

of goopy black bitumen become two million barrels a day.
Soon they will double production. Soon the tankers will cruise

through the Salish Sea, where the orcas spend half the year.
Something about a Goldilocks planet circling its sun

the same distance as Earth orbits hers, only this planet's
1,400 light years away. Only this planet could be gaseous.

Those, if there, if alive, dutifully drilling in their equivalent
of the Arctic, marveling at the rapid melting of permafrost.

(Something about irony. Something about imagination).
Something about not knowing, as in Kinnell's poem,

the thing *unbidden*, like *one-syllabled lumps*, like *toad*,
like *moth*, like *life as we knew it*. Something about

acceleration, a measurement called the albedo, percentage
of light absorbed by the Earth: none, a little, a lot. So many

somethings, unfrozen or charred, emitting methane
and carbon—trees and grass turned against us.

Snow repels light; the widening oceans absorb it exponentially
(is that what he said?). *Locked in a post-glacial warming trend,*

said the woman who said she'd pick up my daughter, running late,
having already been across town and back before the sun rose over

the Cascade crest. Can't pick berries, not think of *squinched,*
same sharp pain as the girl in New Jersey I was, same leaning in.

Something about the bees not seeming to mind a lack
of discernible breeze, grass the color of sand, no trace

of green. Not late September but early August, the news
not icy but the warmest summer on record; we've surpassed

1958. He said he splurged well but when asked said he wasn't
a nature poet. Humans, he said, have taken over, have disregarded

numbers like 1.7 billion, age of the oldest of Death Valley's rocks,
forgotten the cultural lives of crows. Something about the ice

not caring. The sun. The parts per million of carbon
not giving a damn: they just keep going up.

If We Were Not So Single-Minded

Pablo Neruda

How they make it look easy: the slow traffic of cormorants.
The lounging calm of a lion like abandoned property.
Meanwhile, humans and their hammerstones.

What's hidden is how we're more like dandelions than dragonflies.
What's hidden is where the music goes once it escapes
its strings, its hollow avenues of sound.

How far has the first note traveled since the first note escaped
the mouth of a siamang or whale? Maybe when we die
we get to travel where the music lives,

the music of every heavy metal band, the melodies of our ancestors,
the refrains of raptors. There must be a way to save us
from ourselves. There must be a cloud or shroud

or asteroid belt, a minor planet that can knock us back
onto a non-deadly trajectory. Neruda suggests
we should all stop. Imagine if we all

did, walked out into late afternoon light to greet
small pink blooms on a gnarled tree
in a yard that could be anyone's.

Out of Eden

and into my front yard. Kentucky Blue.
Into knowing each year the Hyundai
parked beside my bed of kale melts
a block of Arctic ice the size of a school bus.
I want to kneel right here beside the sumac,
its leaves the sharp red tongues of dragons,
want to hug the boy bicycling down Hudson,
his wild and gorgeous hair. Praise a city
for cleaning grease and oil from sewers
so they don't end up in the Sound.
That I can still find something lifting,
like the heavy geese above me heading south
as I yank a few weeds. Like the dew
that makes the webs in the ivy easier to see.

At the Bottom of the World

At the bottom of the world, two miles
below sea level, tubeworms boggle scientists,

clustered near hydrothermal vents, thriving on
deadly hydrogen sulfide. Here, too, the vampire fish,

echinoderms, giant isopods that won't ever
feel the sun's heat on their pill-buggy backs, the glow

gulls need at least a glimpse of so they can
hassle a minnow from a pelican's pouch.

It's a great place to contemplate
Jesus and his myriad miracles—walking on water,

kersplatting the need for crutches, sidling up with
long-nosed chimeras, one swish of a dorsal fin fatal.

Michelangelo would feel at home here, for
no one who's seen a coffin fish doubts God

opened his palm, pointed a finger, ap-
pointed Adam Creature of Shame, Creature of

Questions More Luminous than Any Star.
Remember when we looked to the heavens,

saw only the campfires of not-so-distant neighbors?
Today it's HD 179949 and Gliese 581D—

unimaginably distant constellations and dwarfs, orbs
vying for habitability. Down here, it's hardly less showy

where lanthanum and neodymium bubble from mud holes, where
xenon nestles in deep-sea basalt. Bless it all, and bless us too,

young and old, bizarre and undeserving, all
zapped with the mystery of the sacred.

Soul Reckoning

I'm skimming through a book called *Spook:*
Science Tackles the Afterlife.
It turns out gravity

doesn't hold for souls; they drift, like loons
between dives, into eternity, along
with NASA's detritus—urine bags,

miles of copper wire, a chunk of Apollo 12. Knowing this,
I leave behind five decades' worth of fear
that I'll die and no longer *be,*

that consciousness must be contained in a body.
To get here, I said goodbye to guilt, to regret
I missed the funeral,

to wishing I was there when my second cousin Timmy
found the lost key to Uncle John's truck,
then sped off to lower her casket

into a hole, to chime or chip in about the words to engrave
on her marking stone. On all of this I punted,
from all of this I'd excused myself

on account of a plane ride, the scepter of unmasked sobbing,
singing, cookies and punch. To get here, I took a road
rutted with limestone creatures

laid down in a shallow sea. My travel speed is two inches per year,
same as the moon from Earth. A little less bound by gravity.
A little more free.

Nature Never Needs Recharging,

its network never goes down, including here
at Big Rock Elementary, where I tell my students

we're putting on our coats, going outside
in search of images. Past the manicured playground,

past the bleachers, toward and beyond the fence,
Cody running past me, yelling *slug eggs!*

I point to trees and flowers and birds, say their names:
Red alder. Licorice fern. Douglas fir. Together,

we listen to the *Oak-a-REE* of a red-winged blackbird,
the raspy call of a red-tail soaring overhead.

Together, we touch the bark of a Hawthorn tree,
admire the fat buds of a big-leaf maple.

I tell each child to choose an artifact,
bring it to their desks. Back inside,

Carmen keeps asking, what's this called?
It's blackberry, I say. Smiling, she twirls it. Sniffs.

An ant hitched a ride on Simon's pinecone.
Tyler wants to know how to tell an eagle

from an osprey. Peyton asks if it's okay
he chose a dead leaf. The room is alive

with rocks and twigs. The poems they write
are crawling with spiders, wiry like lichen,

swaying with catkins, quiet like clumps of moss.
Nature has no username or password. Its signal

is always strong. Just stand in a field,
let your shoes get wet in knee-deep grass.

When I Realized Everything Had Been Said

about soft tissues and hard luck, about honey
and the spoon, deflagrations and worthlessness,
I stare out at a pink late-afternoon sky,
at crows flying south and east

toward what? All day workmen next door
jack hammering a foundation, and now the crows
are heading somewhere milk-soft and unfrantic,
somewhere blastless and lacking affliction,

somewhere they could be dilemma-less,
where the night would be the opposite
of barrage – what would that be like?
The crows growing smaller and smaller

until they're a mess of busy gnats,
out of sight. Where did they go?
Where does the spirit go after a body is char?
I could say they disappeared, but I know they did not.

Not Sitting in a Lifeguard Chair on a Random Day in June

will not be one of my regrets, nor will not waking
at 5:45 am for Zoom yoga, Reese's
reverse half-moons.

Listening to Enchilada lick the empty can
of turkey supreme erased my remorse
for leaving it on the kitchen counter,

that I didn't spring from half pigeon to ruin her bliss.
Nor will I be sorry I did not partake of the extra-
thick cut bacon BLT with extra mayo,

two extra handfuls of potato chips. Never have I been sorry
when I've wakened early. Always it has led me
to cow-dotted hills, a sky resembling a Rothko.

Like a kayaker riding the chop he assumed would be smooth,
I don't regret the disastrous gaffes, like the one
where I tried to pee beside the trail,

post-holed into a snowdrift and twisted my knee.
For two weeks I limped, but all I recall
is the barred owl perched on a fence,

its incessant *who cooks for yooooooouuu?*
Whatever you might end up wishing you'd done—
sampled deep-fried butter at the State Fair, raised ferrets,

swam across your local channel or lake, don't waste a minute
talking yourself out of it: if ever there was a time
to take up the cello,

this is it. Be that person who, when their partner wakes at dawn,
sniffs the lingering scent of coffee and sausage, finds a note:
Cascade Pass by way of Horseshoe Basin. Home after dark.

That Summer

we didn't have enough tomatoes for caprese.
Our sun-gold summer a bust. A friend said
your plants are dying; he was right.

The leaves like maps chewed by moths.
We resigned ourselves to heirlooms
from the store. Delicious, I said,

almost as good, almost a summer but also not—
the floating docks never pulled by a tug
to their beaches, the diving boards

in storage, as if all summer it was winter, a strange season
of heat-struck hydrangeas, an abundance of hunger
for touch, of opening doors, inviting in

the kinglets and wrens. Autumn: here it comes, here it opens
into falling, into steely blankets, into greening grass
where summer had been a *Keep Moving* sign,

a warning: *Crowded Parks Lead to Closed Parks.*
Not a single tomato to can. Something
went wrong in the garden.

Now, the maples are catching fire. Frost overtakes the mud.
I put away my flower-print skirts, turn my attention
to the flames.

When My Father Whistled It Meant Come Home,

come home to the front door with its red peeling paint,
to a man who knew our planet had lost nearly all
its species five times before, knew

there were things much worse than hundreds of gigatons
of CO_2 releasing into the air, things like The Great
Dying, which left our planet treeless

for 10 million years, knew that in the Eocene the Arctic
was a steamy swamp with giant tortoises
and flying lemurs, the ocean

a tepid 76 degrees, that soon enough our world would replicate
that time, carbon's non-human-induced spike
collapsing the West Antarctic Ice Shelf,

melting all of Greenland, drowning Florida, the Nile Delta, Bangladesh.
Knew it didn't matter how much coal and oil we burned,
that between 100,00 and 400,000 years from now

the Empire State Building and Statue of Liberty would be buried under ice.
When my dad whistled, he meant it: come home,
but Dad, but Dad we have a choice;

we can burn it all and go extinct or harness sun and wind,
so you and your kind can keep asking what day it is,
mispronounce Tuesday as *Toozdee,* complain

about the price of cauliflower, run your torsion machines,
twist your polymers, dip hotdogs into liquid nitrogen,
throw them at a wall so they shatter

for school children who need to laugh before they can learn.
Oh, Dad. I want a planet that isn't 95 degrees and rising,
want it to be habitable not only for lizards and birds

but the ones who kiteboard and windsurf, who lie in hammocks and read,
a place where a father can raise two pinkies to his lips, release a whistle
that says the evening's over—it's time to come home.

We Are All Magnificent

like the tail of a fox. We are all the nothingness
after a white dwarf dies. We are all the stiff wind
and the slack, like eaglets, even when we feel itchy

in a blue taffeta dress, even when we walk toward a hollow
where a river otter once roamed. We are all a glassy lake,
a few ripples, a kingfisher in a bare tree. We say goodbye

to the bleeding hearts, welcome poppies the color of yolks.
As the goldfinches molt from ashen to lemon, we are the lemons,
the wonder of citrus. And we are the crystals inside snow,

the tea leaves at the bottom of the cup. In our bones, bits of stars.
We explode like spent suns, like planets dying and being reborn.
We are the heron touching down in the middle of a city,

on a little patch of grass near West 19th, a tad stunned
but figuring out where to find fish. And when we're not
fishing we're dancing to the Beach Boys or Lana Del Rey,

Snoop Dogg or Janelle Monáe. Each of us is a book
of ice that was once a comet, a paperback of Oort Cloud.
Some of us live under a mossy roof, some where monsoons

soak adobe brick. We're mostly hydrogen and oxygen.
Sometimes we're less garbage gyre, more string quartet.
Sometimes we take a catastrophe, turn it into a kiss.

Everything Ends

but so what. In the sultry nights of August, I'll unravel –
wanna join me? We can pant ourselves pantless,

share a double brushfire on the raucous. Together
we can close the book on the Uncertainty Principle,

load up, unwaning, at Wingstop, discuss the sorrow
of burned beaks. Free the crows, you say, and I raise

a toast to a small, uninhabited island, a boisterous whale,
a purity stone, a planet without smoke. Because pleasure

counts big time. Because days spent in a tender mess
are unrecoverable. Naked and floodlit, cocooned

in the opposite of random, we remind ourselves
of the importance of seaweed and seasons,

of each and every bacteria, how we're more
microbe than human. If everything ends,

why are you sharpening your sorrow,
running to catch the discomfort.

Acknowledgments

Gracious thanks to the editors of the following magazines for publishing these poems, some in slightly different versions.

Adirondack Review: "Not Sitting in a Lifeguard Chair on a Random Day in June"
Alaska Quarterly Review: "Oh, Autolysis"
Arkansas International: "When My Father Whistled It Meant Come Home"
Atticus Review: "If We Were Not So Single-Minded"
Bellevue Literary Review: "Letter to a Dead Mother"
Bellingham Review: "Vast"
Bracken: "Mirabile dictu" and "Gray"
Briar Cliff Review: "The Race to Save the Planet's Inhabitants"
The Carolina Quarterly: "2019" and "In the Late Anthropocene"
Cimarron Review: "At the Prairie Creek Campground, Redwood National Park"
Cincinnati Review: "During the Cretaceous Our Country Was Divided for Sixty-Million Years"
The Cortland Review: "December: Summer Lake"
Cutthroat: "Letter to a Post-Apocalyptic Cockroach"
diode poetry journal: "My Watch Face"
EcoTheo Review: "At the Bottom of the World"
Four Way Review: "I Am Writing a Letter"
The Fourth River: "After Apple Picking, Late Anthropocene"
Green Mountains Review: "At the Naples Botanical Garden"
The Ilanot Review: "Leading a Nature Walk on 23rd and Yesler" and "Nature Never Needs Recharging"
Image: "When We Say It's the Little Things"
Indianapolis Review: "What Can I Tell You"
JuxtaProse: "My Son Has a Girlfriend and Other Things I Can't Tell My Mother"
Los Angeles Review: "You're Surrounded"
The Missouri Review: "Two Hundred Million Years Ago This Was All a Sea Floor,"
Moist: "Postcard from the Everglades"
Moria: "Poppy Love"
Night Heron Barks: "Time and Distance"
North American Review: "What is Sad? What is Beautiful? What is Apocalyptic?" and "Failed Attempt at Mythmaking"
Notre Dame Review: "Pine Siskins" and "What They Said"
Ovenbird Poetry: "When My Brother Texts *You Guys Have a Weapon?*"

Poetry Northwest: "Love Song for the Anthropocene"
Presence: "Soul Reckoning" and "We Are All Magnificent"
Psaltery & Lyre: "Poem Written One Hundred Yards from My Mother's Grave"
River Mouth Review: "The Prize Bird of Friday Harbor"
Rust & Moth: "When I Realized Everything Had Been Said"
Salamander: "Today"
Scoundrel Time: "Just before 25 fourth graders crouched beneath a table
      to be instructed on the imperatives of silence and calm,"
Sheila-Na-Gig: "Why I Love to Garden in My Front Yard" and "On the Outside"
The Sextant Review: "Driving through Detroit, Oregon on Highway 22"
Sierra magazine: "No Rain"
South Dakota Review: "At the Burke Museum"
Split Rock Review: "Once,"
Tahoma Literary Review: "Grief Is a Planet" and "At the Wild Horses Monument in
      Vantage, Washington,"
Tar River Poetry: "Letter to My Grandchildren"
Terrain: "My Daughter Is Drawn to Blue Flax"
THRUSH: "Out of Eden" and "That Summer"
Trampoline Poetry: "*Planet Killer* asteroid found hiding in sun's glare may one day hit
      Earth"
Waxwing: "Surpassing Danger"
Zocalo Public Square: "Everything Ends"

I am grateful to the editors for including my poems in the following anthologies:

"Self-Portrait as Southern Resident Orca" appears in *The Ecopoetry Anthology:
      Volume II*, co-edited by Ann Fisher-Wirth and Laura-Gray Street,
      forthcoming from Trinity University Press, 2025.
"My Daughter Is Drawn to Blue Flax," appears in the second edition of
      Environmental and Nature Writing, co-edited by Joe Wilkins and Sean
      Prentiss, Bloomsbury Press, 2024.
"Letter to a Post-Apocalyptic Cockroach" appears in *Dear Human at the Edge of
      Time: Poems on Climate Change in the United States.* Luisa A. Igloria, Aileen
      Cassinetto, Dr. Jeremy Hoffman, co-editors. Paloma Press. 2023.
"In a Font Called Avenir Book" appears in *I Sing the Salmon Home: Poems from
      Washington State.* Rena Priest, Editor. Empty Bowl Press. Chimacum. WA,
      2023.

"To the Ten-Lined Ground Beetle" appears in *Cascadia Field Guide: Art, Ecology, Poetry*. Elizabeth Bradfield, Derek Sheffield, and C. Marie Fuhrman, eds. Mountaineers Books, 2023.

"Self-Portrait as South Resident Orca" appears in *For the Love of Orcas: An Anthology,* co-edited by Andrew Shattuck McBride and Jill McCabe Johnson, Wandering Aengus Press, 2019.

I am most grateful to Playa at Summer Lake for its residency program, where this book began to take shape in the fall of 2019. I also am deeply indebted to the Centrum Residency Program at Ford Worden State Park and the University of Washington's Helen Riaboff Whiteley Center for quiet time in beautiful natural surroundings, where I wrote and revised many of these poems.

This collection would not exist without the writing prompts and positive feedback provided during a two-year pandemic stint of the Thursday Night Poetry Club, a weekly Zoom event with stellar poets Kelli Russell Agodon and Ronda Piszk Broach.

Buckets of gratitude to Michele Battiste for her editorial suggestions that helped me improve this manuscript.

I am gratefully honored to have been a member of the Swift Kickers poetry critique group, founded by Tina Schumann, and including members Jed Myers and Heidi Seaborn. Without their astute and exacting editorial advice, many of these poems would remain unfinished.

I am overjoyed to have been given permission to use David Hytone's painting "Aoyf Papir" as the cover art for this book. Tieton River waves of gratitude, Hytone!

Hats off to Jodi Miller-Hunter, for her astute and aesthetically pleasing book design skills.

Huge and heartfelt thanks to Christopher Howell for choosing to publish this manuscript in Lynx House Press's Pacific Northwest Poetry Series, and to Series Editor Linda Bierds for her poetic and editorial expertise.

About the Author

photo: Langdon Cook

Martha Silano's previous poetry collections include *Gravity Assist* (2019), *Reckless Lovely* (2014), and *The Little Office of the Immaculate Conception* (2011), winner of the Saturnalia Books Poetry Prize and a Washington State Book Award finalist, all published by Saturnalia Books. She is also co-author, with Kelli Russell Agodon, of *The Daily Poet: Day-by-Day Prompts for Your Writing Practice* (Two Sylvias Press, 2013). Martha's poems have appeared in *Poetry, Paris Review, American Poetry Review, Kenyon Review, The Missouri Review*, and in many anthologies, including *Cascadia Field Guide: Art, Ecology, Poetry* (Mountaineers Books, 2023), *Dear America: Letters of Hope, Habitat, Defiance, and Democracy* (Trinity University Press, 2019), and *The Best American Poetry* series (Norton, 2009). Awards include *North American Review*'s James Hearst Poetry Prize and *The Cincinnati Review*'s Robert and Adele Schiff Poetry Prize.

Also by Martha Silano

Gravity Assist (Saturnalia Books, 2019)
Reckless Lovely (Saturnalia Books, 2014)
What the Truth Tastes Like, expanded second edition (Two Sylvias Press, 2015)
The Daily Poet: Day-by-Day Prompts for Your Writing Practice
(Two Sylvias Press, 2013).
The Little Office of the Immaculate Conception (Saturnalia Books, 2011)
Blue Positive (Steel Toe Books, 2006).
What the Truth Tastes Like (Nightshade Press, 1998)